FLORA OF TROPICAL EAST AFRICA

———

HYPOXIDACEAE

J. Wiland-Szymańska[1] & I. Nordal[2]

Perennial herbs with subterranean corms or rhizomes with contractile roots. Leaves in basal, rosulate or tristichous rosette, sessile or pseudopetiolate, ± sheathing at the base, the outermost often reduced to cataphylls; lamina erect or arcuate, linear to lanceolate, ± prominently parallel-veined, often V-shaped in cross-section, or terete, sometimes plicate, pubescent, rarely glabrous. Inflorescences 1-several, racemose, corymbose, spicate or capitate, rarely only one-flowered; scapes flattened or rarely terete. Flowers bisexual, rarely andro-dioecious, actinomorphic, usually trimerous; tepals subequal, most often yellow, free. Stamens 6, filaments short, arising from the base of the tepals; anthers opening latrorsely or introrsely by longitudinal slits (apical pores in *Molineria*). Ovary inferior, 3-locular with axile placentation, ovules few to many. Fruits indehiscent (*Curculigo* and *Molineria*) or dehiscent (*Hypoxis*), opening by a transverse slit (circumscissile) or longitudinal slits (loculicidal). Seeds dark brown or black, ellipsoid to globose.

A tropical to subtropical family of 9 genera, mainly in the southern hemisphere. The two genera known in the wild from East Africa, *Hypoxis* and *Curculigo*, are pantropical. *Molineria*, cultivated as an ornamental in East Africa, is of Asiatic origin. Four genera are endemic to South Africa, one is shared between South Africa and Australia, and one is endemic to the Seychelles.

Corms of some of the species of Hypoxidaceae are used for medicinal purposes (said to be functional against AIDS, cancer, and gastric problems) and are in great demand at the moment. Especially species of *Hypoxis*, of which the corms are called "African potatoes" might be endangered due to this.

1. Leaf width up to 7 cm; flowers solitary to 25 in a racemose, corymbose or spicate inflorescence; indigenous plants 2
   Leaf width larger than 7.5 cm; flowers more than 50 in dense heads; cultivated . . . . **Molineria**
2. Ovary beak absent; fruits displayed above ground, dry-walled capsules with vertical or horizontal split . . . . 1. **Hypoxis**
   Ovary with a long or short beak separating the tepals from the ovary; fruits subterranean, indehiscent, with a succulent to membranous wall . . . . 2. **Curculigo**

Molineria capitulata (*Lour.*) *Herbert* has been cultivated in our area. It is originally from tropical Asia, Philippines and North Australia; it has also been cultivated in Africa and South America.

Perennial herb with leaves to 1 m long; lamina lanceolate, plicate. Inflorescence racemose, capitulate, deflexed, ovoid or globose, 2.5–5 cm diameter. Flowers yellow; rostrum about 3–4 mm long; perianth segments ovate, 6–8 mm long, acute; anthers lanceolate, 3–4 mm long, filaments short; ovary turbinate, 6–8 mm, densely villose, with about 10 ovules. Fruit indehiscent, globose, 6–8 mm. Seeds numerous, black, ovoid, 1.5 mm diameter; testa with irregular, coarse striations.

Kenya. Trans-Nzoia District: Kitale, 1963, *Horsfall* H257/63!
Tanzania. Arusha District: Arusha, 19 Mar. 2001, *Wiland & Mboya* 175! & Ilkyushin, 2 June 1991, *Kalema & Raphael* 218!; ibid., 19 Mar. 2001, *Wiland & Mboya* 176!

1. HYPOXIS

L., Syst. Nat., ed. 10, 2: 986 (1759); Baker in J.L.S. 17: 98–119 (1878) & in F.T.A. 7: 377; Nel in E.J. 51: 259–286, 301–340 (1914); Geerinck in B.J.B.B. 39: 72–80 (1969) & in F.C.B.: 4–9 (1971); Nordal et al. in Nordic J. Bot. 5: 15–30 (1985); Nordal & Iversen in Fl. Cameroun 30: 34–47 (1987) Champluvier in Fl. Rwanda 4: 81–84 (1987); Nordal in Fl. Eth. 6: 86 (1997) & in Kubitzki, Fam. Gen. Vasc. Pl. 3: 292 (1998); Nordal & Zimudzi in F.Z. 12 (3): 1–18 (2001); Wiland-Szymańska in Ann. Missouri Bot. Gard. 88: 301–350 (2001)

Corms most often surmounted by a ± dense coat of fibrous leaf remnants and ringed by stout contractile roots usually arranged in an equatorial zone. Leaves basal, sessile, linear to lanceolate, V-shaped in cross-section (conduplicate) or flat to ± terete (sometimes pseudopetiolate – not in our area), sheathing at the base, pubescent along the abaxial midrib and margins, and often also on the lamina, sometimes forming a pseudostem (from the leaf sheaths and cataphylls), outermost leaves often reduced to cataphylls, new leaves produced successively from within older leaf bases; indumentum of white, yellowish to brownish 2-armed (bifurcate) or 3–12-armed (tufted) hairs. Inflorescences produced continuously throughout the season, or only at its beginning, on flattened scapes covered with tufted hairs, particularly in upper parts and on the rhachis. Flowers always supported by small hairy bracts, most often arranged in spicate to corymbose inflorescences, rarely only solitary; perianth segments (tepals) free, acute to obtuse, persistent, yellow adaxially (inside); outer tepals greenish and pubescent abaxially, inner tepals yellowish-green and pubescent only on the abaxial (outside) midrib; stamens equal in length or in the inner whorl shorter than in the outer, with short filaments; anthers latrorse, sagittate at the base, thecae fused or free at the apex. Fruit a capsule with circumscissile (across carpels) or loculicidal (parallel to carpels) dehiscence. Seeds with a distinct micropyle; testa shiny black or dull brownish, smooth or variously papillose; the cuticle of the papillae smooth, or with foldings, most often as wing-like striae radiating from the papilla apex with fine irregular reticulate striae between (Fig. 1, 1–9, p. 4).

A genus of about 50 species, widespread in grassland and wooded grassland in Africa, America, Asia and Australia. Most of them appear to be fire tolerant, some even needing fire to flower.

The genus includes a few diploid species with normal sexual reproduction (as *H. angustifolia* and *H. filiformis*) and several highly polyploid species that reproduce apomictically (asexual seed set, genetically equivalent to cloning). Plants with apomixis always create taxonomical problems, as the variation patterns are difficult to compare with those of sexual species, and species delimitation becomes complicated. Different morphological forms often occur in the same "population" as there is no gene exchange. Some authors have been splitters – and probably largely described clones (e.g. Nel 1914) – others have been lumpers (e.g. Nordal et al. 1985). Here we have tried to compromise. Generally, more field observations (particularly on underground organs) and microscopic observations (particularly on indumentum and seed surfaces) are needed. Adding to the difficulty in identifying taxa is the fact that the plants continue to produce new inflorescences while the leaves develop, thus completely altering the appearance through the season. Also the indumentum density changes through the season: in the same plant young leaves may appear densely pubescent, and old leaves may appear almost glabrous.

When we refer to leaves in the key, this does not apply to the outer cataphylls, which might differ in indumentum and certainly in shape, but to well-developed inner leaves in the rosette.

1. Plants with no or indistinct pseudostem (up to 5 cm long); inflorescences variable 2
   Plants with a 6–15 cm long distinctive pseudostem; inflorescence spiciform, flowers alternating along the rhachis; corm orange or yellow inside . . . .   15. *H. rigidula*
2. Leaves filiform to grass-like, up to 6 mm wide 3
   Leaves wider than 7 mm 8
3. Seeds brownish, dull, due to cuticular folding (Fig. 1: 3, 6, 9) 4
   Seeds black, glossy without cuticular folding (Fig. 1: 1, 2, 4 5, 7, 8) 5
4. Outer leaf remnants membranous; pedicels more than 10 mm long; perianth 10–18 mm diameter when open; capsule opens with longitudinal slits . . . .   1. *H. angustifolia*
   Outer leaf remnants fibrous; pedicels less than 6 mm long; perianth 18–31 mm when open; opening of capsule circumscissile . . . .   2. *H. schimperi*
5. Seed testa not papillose, periclinal walls flat (Fig. 1, as in 1)   3. *H. malaissei*
   Seed testa papillose, periclinal walls convex (Fig. 1, as in 2, 5) 6
6. Leaves up to 4 mm wide; corms white inside 7
   Leaves usually more than 4 mm wide; corms yellow inside   13. *H. nyasica*
7. Remnant of old leaf bases membranous; scapes bend down after flowering . . . .   4. *H. kilimanjarica*
   Remnant of old leaves stiff and prominently fibrous; scapes erect after flowering . . . .   5. *H. filiformis*
8. Seed testa brownish dull, due to cuticular folding (Fig. 1, as in 3, 6, 9) 9
   Seeds black, often glossy, rarely with cuticular foldings and dull 11
9. Leaves glabrous except for tufted hairs along margin and midrib; inflorescence starts flowering from the top downwards (basipetalous); corm white inside; flowers often appearing before the leaves are developed . . . . . .   6. *H. goetzei*
   Leaves with indument, on at least one side of the lamina; inflorescence starts flowering from the base upwards (acropetalous) or irregularly; corm white or yellowish inside; flowers appearing after the leaves are developed 10
10. Leaves with thick tomentose indumentum on both sides of the lamina, white in the field, drying golden or golden-red . . . .   7. *H. bampsiana*
    Leaves villose, pilose-pubescent on the lower side of the lamina, glabrous or with only a few hairs on the upper side, always white . . . .   8. *H. polystachya*
11. Lamina glabrous except margin and midrib (very young leaves may have some hairs) 12
    Lamina with hairs at least on one surface 15
12. Inflorescence spiciform, flowers alternating along the rhachis, upper sessile, lower with pedicels to 8 mm long; plants drying red-brown. . . . .   9. *H. galpinii*
    Inflorescence racemose, flowers opposite along rhachis, at least lower distinctly pedicellate up to 35 mm; plants drying yellowish 13

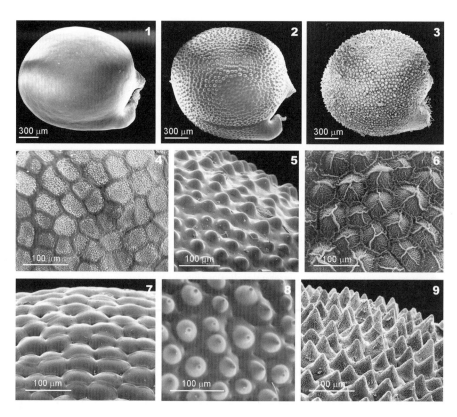

FIG. 1. *HYPOXIS* — **1**, seed of *H. gregoriana*; **2**, seed of *H. galpinii*; **3**, seed of *H. goetzei*; **4**, seed testa of *H. gregoriana*; **5**, seed testa of *H. galpinii*; **6**, seed testa of *H. angustifolia*; **7**, seed testa of *H. urceolata*; **8**, seed testa of *H. fischeri*; **9**, seed testa of *H. polystachya*. 1 from *Simon et al.* 790; 2 & 5 from *Wiland & Mboya* 57; 3 from *Wiland & Mboya* 97; 4 from *Morna Hale* 57; 6 from *Bjørnstad* 561; 7 from *Katende* K2126; 8 from *Lye* 2079; 9 from *Wiland & Mboya* 159A.

13. Leaf veins equal in width, tunic fibers copius, matted, red-brown; number of flowers 7–13 . . . . 10. *H. obtusa*
Leaf veins unequal in width, tunic fibrous, black; number of flowers 4–7 14

14. Leaves 3–8(–20) mm wide, ± erect; seed testa with periclinal cell wall convex; southern Tanzania . . . . 13. *H. nyasica*
Leaves (8–)12–20 mm wide, often reflexed; seed testa with periclinal walls ± flat; northern Tanzania, Kenya and Uganda . . . . 14. *H. urceolata*

15. Hairs on the lamina all tufted 16
Hairs on the lamina (except margin and midrib) bifurcate 18

16. Seeds with cuticular folding; leaves villose, pilose-pubescent on the lower side of the lamina, glabrous or with only a few hairs on the upper side . . . . 8. *H. polystachya*
Seeds without cuticular folding; indumentum on both sides of lamina 17

17. Leaves erect with dense indumentum; inflorescence spike-like, number of flowers 6–12; pedicels 0.2–2 cm . . . . . . 11. *H. fischeri*
Leaves recurved with sparse indumentum; inflorescence corymbose, number of flowers 2–4; pedicels 2–3 cm . . 12. *H. gregoriana*

1. **Hypoxis angustifolia** *Lam.* in Encycl. Méth. Bot. 3: 182 (1789); Baker in J.L.S. 17: 111 (1878); Durand & Schinz in Consp. Fl. Afr. 5: 231 (1895); Baker in F.T.A. 7: 378 (1898); Nel in E.J. 51: 303 (1914); Cufodontis in Miss. Biol. Borana, Racc. Bot. 4: 328 (1939); Jex-Blake in W.F.K.: 130 (1948); U.O.P.Z.: 302, fig. (1949); F.P.S. 3: 306 (1956); F.P.U: 212 (1962); Hepper in F.W.T.A. ed. 2, 3: 172 (1968); E.P.A.: 1577 (1971); Geerinck, F.A.C.: 5, fig. pro parte (1971); Wickens, Fl. of Jebel Marra: 160, Map 165 (1976); Cribb & Leedal in Mount. Fl. S. Tanz.: 167, t. 46B (1982); Nordal et al. in Nordic J. Bot. 5: 24 (1985); Champluvier in Fl. Rwanda 4: 84 (1987); U.K.W.F., ed. 2: 313, t. 142 (1994); Thulin in Fl. Somalia 4: 31 (1995); Zimudzi in Kirkia 16:15 (1996); Nordal in Fl. Eth. 6: 87 (1997); Wood in Handb. Fl. Yemen: 405 (1997); Nordal & Zimudzi in F.Z. 12, 3: 1 (2001); Wiland-Szymańska in Ann. Missouri Bot. Gard. 88: 309 (2001) & in Novon 12: 142 (2002). Type: Mauritius, *Commerson* s.n. (P!, holo.)

Slender herbs up to 53 cm tall. Corm (sub-)globose, 0.8–2 cm diameter, white inside, outer leaf remnants membranous. Leaves forming a whitish pseudostem up to 5 cm long, linear, grass-like, ± erect, up to ± 50 cm × 1.5–10 mm, linear, sparsely appressed-pilose with white to golden soft, 2-branched, up to 2.5 mm long hairs, mostly on the margins and midrib abaxially (outside); 5–13 veins of unequal size. Inflorescences 1–6, with slender scapes, 5–20 cm long, ± 1 mm wide. Flowers 1–4(–7) in lax corymbs; pedicels slender, 12–40 mm long; bracts 5–17 mm long, filiform to linear; tepals 5–8 × 2–3(–4) mm, inner slightly wider than outer, narrowly elliptic to ovate, thinly pilose abaxially (outside), bright yellow inside, green outside, sometimes red-keeled; stamens usually unequal, outer 2.2–4.5 mm long with filaments 1.3–3, inner 1.5–3.5 mm long with subulate filaments 1–2 mm long, anthers 1–3.2 mm long, split at the apex; ovary obconical, 2–5 mm × 1–3 mm, style 0.5–3 mm long, stigma 0.6–2.6 mm long, both variable in shape. Capsule 3.5–12(–15) × 2–4 mm, turbinate-cylindrical, thinly pilose, loculicidal.

var. **luzuloides** (*Robyns & Tournay*) *Wiland* in Novon 12:148 (2002). Type: Congo-Kinshasa, Kivu, Tschambi, *de Witte* 1130 (BR!, holo.)

Seeds dark brown, subglobose, 0.8–1.2 mm diameter; testa papillose, each papilla with a minutely wrinkled cuticle and 3–4 longitudinal narrow wing-like ribs (as in Fig. 1: 6, p. 4).

UGANDA. Ankole District: Ruizi River, 15 May 1950, *Jarrett* 55!; Busoga District: Butembe Bunya, on S slopes of Wanyange Hill 8 km E of Jinja, 22 Aug. 1952, *G.H.S. Wood* 354!; Mengo District: Wakyato-Luwero, 21 Apr. 1956, *Langdale-Brown* 2067!
KENYA. Northern Frontier District: Moyale, 16 Apr. 1952, *Gillett* 12799!; Kisumu-Londiani District: Tinderet Forest Reserve, Camp 4, 29 June 1949, *Maas Geesteranus* 5220!; Kwale District: Shimba Hills, Pengo Hill area, 27 Mar. 1968, *Magogo & Glover* 498!
TANZANIA. Moshi District: 16 km on Moshi–Arusha road, 15 Dec. 1961, *Polhill & Paulo* 987!; Tanga District: flats by Lwengera River, 4 km E of Korogwe, 16 July 1953, *Drummond & Hemsley* 3335!; Dodoma District: road to Homboro from Great North Road (near Dodoma), 7 Mar. 1966, *McCusker* 98!; Zanzibar, Kizimbani, 20 May 1959, *Faulkner* 2262!
DISTR. **U** 1–4; **K** 1–7; **T** 1–8; **P**, **Z**; widespread in tropical and southern Africa; Madagascar and Yemen

HAB. Grassland, swampy grassland, scattered tree grassland, burnt and secondary grassland, open woodland, forest margins, shallow soil over rock; 0–3000 m

SYN. *H. luzuloides* Robyns & Tournay in B.J.B.B. 25: 254 (1955); Robyns & Tournay in F.P.N.A. 3: 390, t. 53 (1955)

NOTE. It has turned out that the seed structure is different in the type specimen of *H. angustifolia* from Mauritius, in lacking the typical cuticular folding that is found on the seeds of the African mainland plants (Fig. 1: 3 & 6, p. 4), and therefore is referred to a separate variety. See Wiland-Szymańska & Adamski in Novon 12: 142–151 (2002).

2. **Hypoxis schimperi** *Baker* in J.L.S. 17: 110 (1878) & in F.T.A. 7: 378 (1898); Nel in E.J. 51: 305 (1914); Cufodontis in E.P.A.: 1578 (1971); Nordal in Fl. Eth. 6: 89 (1997). Type: Ethiopia, Gonder (Begemder), *Schimper* 1118 (K!, holo.)

Slender herbs up to 45 cm tall. Corm ± cylindrical, 2–4.5 × 1.5–3 cm, white or greenish-yellow inside. Leaves forming a whitish pseudostem up to 5 cm long, narrowly lanceolate, ± erect, up to ± 45 cm × 1.5–6 mm, glabrous or thinly pilose, mostly on the margins and midrib abaxially (outside) with hairs 2-armed, the arms fine, appressed, ± unequal; only 2 visible veins except the midrib. Inflorescences 1–6, with slender scapes up to 21 cm long, 0.7–1 mm wide. Flowers most often solitary; pedicels 2–6 mm long; bracts 5–15 mm long, filiform to linear-lanceolate; tepals up to ± 12 × 3–4 mm, narrowly elliptic, thinly pilose abaxially (outside); stamens equal, filaments subulate, 1.3–4 mm long, anthers 2.6–3.2 mm long, not split at the apex; ovary obconical, 3 mm long, style 0.5–1.6 mm long, stigma 1.5–2.3 mm long. Capsule 5–10 × 4–5 mm, turbinate-cylindric, thinly pilose, circumscissile. Seeds dark brown, 0.8–1.2 mm diameter, globose; testa papillose, each papilla with a minutely wrinkled cuticle and 3–4 longitudinal narrow wing-like ribs (as in Fig. 1: 3 & 6, p. 4).

KENYA. Trans-Nzoia District: Kitale, 6.5 km on road to Moi's [Hemstead] Bridge, 6 Aug. 1968, *Agnew, Kibe & Mathenge* 10588! & near Kapenguria road 5 km from Kitale, May 1954, *Rayner* 541!; Teita District: NE slope of Yale Mt, 10 Apr. 1966, *Gillett* 17261!
TANZANIA. Mpanda District: Kapapa Camp, 28 Oct. 1959, *Richards* 11621!; Kondoa District: Bereko, 17 Jan. 1974, *Richards & Arasululu* 28737!; Songea District: 6.5 km west of Songea, 6 Jan. 1956, *Milne-Redhead & Taylor* 8058!
DISTR. **K** 3, 4?, 5?, 7; **T** 2?, 4, 5, 7, 8; Ethiopia, Zambia, Zimbabwe
HAB. Grassland, swampy grassland; 900–2400(–3400?) m

SYN. *H. macrocarpa* Holt & Staubo in Nord. J. Bot. 5: 25 (1985). Type: Tanzania, Ufipa District, Sumbawanga, near Mpui, *Richards* 8768 (K!, holo.)
H. cuanzensis sensu Zimudzi in Kirkia 16: 15 (1997); Nordal & Zimudzi in F.Z. 12: 7 (2001), non *H. cuanzensis* Baker

NOTE. In F.Z. this taxon was named *H. cuanzensis* Baker. The type of *H. cuanzensis* (*Welwitsch* 4056) is from Angola. These two taxa are closely related, but they differ somewhat in seed cuticular folding and petal shape, *H. cuanzensis* being somewhat narrower, and in leaf venation, veins being more prominent in *H. cuanzensis*. It is uncertain whether the two taxa deserve taxonomic recognition at species level. The plants from East Africa (and in the F.Z. area) are very similar to the type of *H. schimperi* from Ethiopia, and should be named accordingly. Plants with seed cuticle folding identical to *H. schimperi*, but with somewhat narrower leaves (and often with four tepals) were named *H. monanthos* Baker for in Central Africa (see Wiland-Szymańska in Ann. Missouri Bot. Gard. 88: 333, 2001). Relations between these three taxa need further investigation.

Three East African specimens appear to deviate in testa structure (*Moreau* 80 from **T** 2, *Gatheri, Mungai & Kibui* 79/84 from **K** 4, *Davis* 81 from **K** 5), lacking the typical cuticular folding, which might be due to seeds being unripe. Two of them are from higher altitudes (2500–3460 m) than the rest of specimens. This also needs further investigation.

3. **Hypoxis malaissei** *Wiland* in Fragm. Florist. Geobot. 42: 418, fig. 6 (1997); Wiland-Szymańska in Ann. Missouri Bot. Gard. 88: 329, fig. 14 A & B (2001). Type: Congo-Kinshasa, Shaba, 28 km NE of Lubumbashi, *Malaisse* 7403 (BR!, holo.)

Slender herbs, 15–36 cm tall. Corm globose, ± 1 cm diameter, surmounted by membranous and fibrous black old leaf-bases. Leaves linear, 7.5–36 × 2.5–4 mm, with white 2-branched, hairs; 7–9 veins of unequal size, two of them as thick as the midrib. Inflorescences 1–5, 9.5–20 cm long and 1.5 mm wide. Flowers 2–5 in a spicate cyme, sometimes only one flower developing; bracts subulate, 5.5–10 mm long; pedicels 7–12 mm long; tepals elliptic or obovate, acute, 6.5–7.5 × 2.5–3 mm; stamens equal, filaments filiform, 2–2.5 mm long, anthers 2–3 mm long, sagittate at base, thecae fused apically; ovary obconical 3.5 × 2.5 mm, style slender, 1.5–2 mm long, stigma 1 mm, composed of three linear rows of papillae. Capsule obovoid, 5 × 3.5 mm. Seeds ovoid, 2 × 1 mm, testa flat, non-papillose with honeycomb-shaped cells (as in Fig. 1: 1, p. 4).

TANZANIA. Iringa District: Iringa, 1 Jan. 1971, *Bjørnstad* 401!
DISTR. **T** 7; Congo-Kinshasa
HAB. Degraded miombo woodland; ± 2000 m

NOTE. This is a very rare taxon known only from the type locality in Upper Katanga and from Iringa. Its testa structure is different from all other *Hypoxis* species studied.

4. **Hypoxis kilimanjarica** *Baker* in F.T.A. 7: 378 (1898); Nel in E.J. 51: 303 (1914); Nordal et al. in Nord. J. Bot. 5: 26 (1985); Champluvier in Fl. Rwanda Sperm. 4: 84, fig. 28: 1A, 1B (1987); U.K.W.F. 2. ed.: 312 (1994); Wiland-Szymańska in Ann. Missouri Bot. Gard. 88: 324–327, fig. 11A, 12A & B (2001). Type: Tanzania, Moshi District, Marangu, Raussi stream, *Volkens* 781 (B!, holo.; K!, BM!, iso.)

Slender herbs up to 20 cm high. Corm subglobose, 6–14 × 6–13 mm, white inside, remnants of old leaves mainly membranous, occasionally with some soft fibers. Leaves narrowly linear, erect-recurvate or prostrate, often spathe-like in basal part, 3.5–22 cm × (1–)2–4 mm, sparsely pilose on margins and midrib below, sometimes also on the blade surface; trichomes 2- to 3-branched; veins 9–13, unequal. Inflorescences 1–4, with slender scapes 2–16 cm long, 0.3–1 mm wide, usually bending downwards after flowering, narrowly winged and ciliate at base. Flowers single, rarely 2; bracts subulate, very narrow, acute, 3–10 mm long; pedicels 2.5–6 mm long; tepals 6, exceptionally 3 or 4, elliptic to ovate, acute, 4–7 × 1.5–3 mm; stamens unequal, outer 3–4 mm with filaments 2–3 mm long, inner 2.5–3.5 mm with subulate filaments 1.5–2 mm long, anthers 1.5–2 mm long, thecae fused and obtuse at apex; ovary obconical, 2–4 mm × 1.5–2 mm; style wider in basal part, 1–2 mm long; stigma pyramidal, obtuse at apex, with three stripes of papillae, 1–2 mm long. Capsule obovoid, 4–8 × 2.5–3.5 mm, almost glabrous; seeds numerous, ovoid, ± 1.5 × 1 mm, black; testa papillose without cuticular folding (as in Fig. 1: 2 & 5, p. 4).

Leaves erect-recurved, flat, thin; scape 4–16 cm long, erect
during anthesis, later recurved . . . .   subsp. *kilimanjarica*
Leaves prostrate, somewhat succulent; scapes up to 4 cm
long, flowers and fruits at ground level . . . .   subsp. *prostrata*

subsp. **kilimanjarica**; *Nordal et al.* in Nord. J. Bot. 5: 27 (1985)

KENYA. Northern Frontier District: Mt Nyiru, 30 Mar. 1995, *Bytebier et al.* 90!; Naivasha District: Aberdare Mts, Kinangop, 2 Apr. 1922, *Fries & Fries* 2698!; Masai District: Olokurto, 13 May 1961, *Glover, Gwynne & Samuel* 951!

TANZANIA. Moshi District: N side of Kibo, 26 Dec. 1932, *Geilinger* 5097!; Lushoto District: W
    Usambaras, 2.5 km NE of Bumbuli Mission on path to Mazumbai, 10 May 1953, *Drummond
    & Hemsley* 2459! & Shagayu Forest Reserve, summit 2.5 km ENE of Shagayu Sawmill, 14 Mar.
    1984, *Borhidi et al.* 84/869!
DISTR. **K** 1, 3, 6; **T** 2, 3; Congo-Kinshasa, Rwanda, Burundi
HAB. Upland grassland; 1650–3200 m

SYN. *H. incisa* Nel in Bot. Jahrb. Syst. 51: 319 (1914); Robyns & Tournay in Fl. Sperm. P.N.A.:
    388 (1955). Type: Tanzania, Lushoto District: West-Usambara, Lutindi, *Liebusch* s.n.
    (B!, syn.)
    H. alpina R.E.Fries in K. Svenska Vetensk.-Akad. Handl. ser. 3, 25: 78 (1948). Type: Kenya,
    Aberdare Mts, Kinangop, *Fries & Fries* 2698 (UPS!, holo.)
    H. angustifolia sensu Geerinck in F.A.C.: 5 (1971) pro parte, *non* Lam.

NOTE. The specimen *Synge* 1229 from Uganda, **U** 2, Kigezi District might represent this taxon,
    but more material is needed.

   subsp. **prostrata** *Holt & Staubo* in Nord. J. Bot. 5: 25 (1985). Type: Kenya, Elgeyo District: N
Cherengani, Chepkotet, *Thulin & Tidigs* 201 (UPS!, holo.; EA, K!, S, iso.)

KENYA. Elgeyo District: N Cherangani Hills, Chepkotet, 12–13 Aug. 1968, *Thulin & Tidigs* 201!;
    Laikipia District: Aberdare range, 3 km SE from Shamata Forest Station, 24 Oct. 1970,
    Mabberley 372!; Nakuru District: 32 km S of Thomsons Falls, 14 Aug. 1952, *Bogdan* 536!
TANZANIA. Kilimanjaro, 28 Jan. 1960, *Rauh* 129! & idem, junction of main trail with track to
    Maundi Crater, 9 Sep. 1993, *Grimshaw* 93/693! & above Mandara Hut, 28 Jan. 1979, *Staubo* 3!
DISTR. **K** 2, 3; **T** 2; not known elsewhere
HAB. Montane grassland, often on wet soil; 2900–3500 m

NOTE. This subspecies might turn out to deserve a specific rank. More comparative studies in
    the field should be undertaken.

   5. **Hypoxis filiformis** *Baker* in J.L.S. 17: 109 (1878) & in F.C. 6: 180 (1896); Nel in
E.J. 51: 305 (1914); Zimudzi in Kirkia 16: 15 (1996); Nordal & Zimudzi in F.Z. 12, 3:
10 t. 12.3.2, fig. B. (2001); Wiland-Szymańska in Ann. Missouri Bot. Gard. 88: 317, fig.
6D (2001). Type: South Africa, Queenstown, *Cooper* 462 (K!, syn.; B! isosyn.) & Natal,
Mohlamba Mts., *Sutherland* s.n. (K, syn.)

   Slender herbs, 20–50 cm tall. Corm vertical, cylindrical to subglobose, 1–4.5 ×
0.5–4 cm, whitish or yellow inside, crowned by stiff, black old leaf remnants. Leaves
forming a pseudostem up to 5 cm long, grass-like, linear, often stiffly erect, broadly
sheathed at the base and narrowing into the blade, 5–61 cm × 1–3 mm, finely pilose-
pubescent with 2-branched ± 3 mm long mostly appressed red-brown to golden
(rarely white) hairs, particularly on the margins and midrib abaxially, glabrescent;
strongly ribbed with 5–11 veins, submarginal veins often thicker than the midrib.
Inflorescences 1–10, with slender scapes up to 57 cm long, 0.5–1 mm wide. Flowers
1–5 in corymbose cymes, first appearing before the leaves; pedicels very different in
length, the lower ones 0–3 mm long, the upper to 20 mm long; bracts linear-
subulate, 2–9 × 1 mm; tepals narrowly elliptic to ovate, up to 0.5–12 × 2–4 mm, acute,
inner somewhat broader than outer; stamens unequal, outer 2.5–5 mm long with
filaments 2–3.5 mm long, inner 1.5–3 mm long with filiform filaments 0.5–1.5 mm
long, anthers 1.2–2.5 mm long, emarginate at apex; ovary obconical, 2–3.5 mm ×
1.5–2.5 mm, style 0–0.5 mm long, stigma 1–2 mm long, composed of 3 lobes, entirely
fused or emarginate. Capsule obovoid to turbinate, 2–5 × 2–3 mm, opening by a
transverse slit. Seeds black and glossy, (sub-)globose, 0.6–1 mm diameter, testa
papillose, the papillae dome-shaped with a smooth shiny cuticle (as in Fig. 1: 2 &
5, p. 4).

UGANDA. Kigezi District: Mgatunga, Oct. 1940, *Eggeling* 4119! & Ingalinge, Mellom Sabinio,
    1933, *Eggeling* 1106!

TANZANIA. Mpanda District: Kapapa Camp, 28 Oct. 1959, *Richards* 11617!; Iringa District: Ruaha National Park, 0.5 km N of Maganwe air strip, 2 Jan. 1973, *Bjørnstad* 2505!; Songea District: Kimarampaka steam, about 12 km W of Songea, 31 Dec. 1955, *Milne-Redhead & Taylor* 8001!
DISTR. U 2; **T** 4, 6, 7, 8; Congo-Kinshasa, Burundi, Angola, Zambia, Malawi, Mozambique, Zimbabwe, Swaziland, Lesotho, South Africa
HAB. Grassland, boggy grassland, dry grassland, burnt grassland; 1050–3100 m

SYN. *H. malosana* Baker in K.B. 1897: 284 (1897) & in F.T.A. 7: 379 (1898); Binns in Fl. Malawi: 53 (1968); Nordal et al. in Nord. J. Bot. 5: 27 (1985); Wiland-Szymańska in Ann. Missouri Bot. Gard. 88: 317, figs. 15A–C, 16 (2001). Type: Malawi, Mount Malosa, near Zomba, *Whyte* s.n. (K!, holo.)
   H. münzneri Nel in E. J. 51: 307 (1914). Type: Tanzania, Kigoma District: Mtembwa, *Fromm* 127a (B!, holo.)

6. **Hypoxis goetzei** *Harms* in E. J. 30: 276 (1901); Nel in E. J. 51: 320 (1914); Nordal et al. in Nord. J. Bot. 5: 25 (1985); U.K.W.F. 2nd ed.: 312. t. 141 (1994); Zimudzi in Kirkia 16: 14 (1996) pro parte; Nordal & Zimudzi in F.Z. 12, 3: 8, t. 12.3.2, fig. A (2001); Wiland-Szymańska in Ann. Missouri Bot. Gard. 88: 319, figs. 5C, D, 8, 13C. (2001). Type: Tanzania, Mbeya District, Unyika, Toola, *Goetze* 1416 (B!, holo.; BM!, BR!, iso.)

Robust herbs up to 40 cm tall. Corm subglobose, white inside, 2–5.7 cm diameter, prominently crowned by black fibrous remnants of old leaves. Leaves vividly green, white in basal part, drying reddish-brown, ± conduplicate, outer arching, inner rigid and twisted when mature, ovate-lanceolate to lanceolate or narrowly elliptic, 11–40(–58) × 2–7 cm, ciliate on the margins and midrib below, glabrous above; trichomes tufted 3–12-armed, up to 2 mm long, red brownish to dark golden; nerves unequal, 27 to 91. Inflorescences 1–10, stout, 8–27 cm long and 5(–8) mm wide, width reducing in steps as pedicels diverge. Flowers usually appearing before the leaves, (4–)6–12(–14) in a spike, opening from top to bottom (basipetalous); pedicels 2–10(–16) mm long; bracts subulate to lanceolate, 7–18 mm long, 4 mm wide, often ciliate on margins, exceeding the pedicels. Tepals ovate to lanceolate, acute, 10.5–21 × 4–8(–9) mm broad, the inner tepals wider than the outer; stamens equal, filaments subulate, 2.5–5 mm long, anthers 4–7(–8) mm long with thecae fused apically; ovary obconical, (3–)6–7(–9) × 3–6(–7) mm, style often tapering towards the base, 2–4 mm long, stigma 2–3 mm long, narrowly pyramidal, composed of three ± free lobes variously covered with papillae. Capsule turbinate-obconical, 4–10 × 3–5(–6) mm, circumscissile. Seeds greyish-brown, 1.6–1.8 mm diameter; testa papillose, each papilla with a minutely wrinkled cuticle and 3–4 longitudinal narrow wing-like ribs (Fig. 1: 3 & 6, p. 4).

KENYA. Trans-Nzoia District: S Cherangani, 15 Feb. 1958, *Symes* 286! & Kitale, Apr. 1962, *Tweedie* 2342! & Aug. 1962, *Tweedie* 2401!
TANZANIA. Ufipa District: Mbizi Forest, 21–26 Nov. 1958, *Napper* 1075!; Iringa District: Magangwe airstrip, 10 Dec. 1970, *Greenway & Kanuri* 14764!; Songea District: about 12 km W of Songea by Kimarampaka stream, 1 Jan. 1956, *Milne-Redhead & Taylor* 8013!
DISTR. **K** 3; **T** 4, 7, 8; Congo-Kinshasa, Zambia, Malawi, Zimbabwe
HAB. Burnt grassland, *Brachystegia* woodland, forest margins; 900–2600 m

SYN. *H. esculenta* De Wild. in F.R. 11: 537 (1913). Type: Congo-Kinshasa, Upper-Katanga, *Hock* s.n. (BR!, holo.)
   H. rubiginosa Nel in E. J. 51: 320 (1914). Type: Tanzania, Ngaka [Mgaka] valley, *Busse* 947 (B!, holo.)

7. **Hypoxis bampsiana** *Wiland* in B.J.B.B. 66: 207, figs. 1 & 20 (1997); Nordal & Zimudzi in F.Z. 12, 3: 8–10 (2001); Wiland-Szymańska in Ann. Missouri Bot. Gard. 88: 312, figs. 4, 5A, B (2001).Type: Congo-Kinshasa, upper Shaba [Katanga], *Lisowski, Malaisse & Symoens* 7653 (POZG!, holo.)

Robust herb up to 45 cm. Corm ovoid, 5.5–6 cm long and 2.8–6 cm wide, white inside, surmounted by the membranous and fibrous relicts of the old leaves. Leaves erect or slightly reflexed, golden or red-golden after drying, rather grey in the field, thick, ovate or lanceolate, (12–)13.5–46(–56) cm long and (8–)14–70(–76) mm wide, cuspidate, slightly keeled, tomentose bifacially or only abaxially, basally glabrous; hairs tufted, 9–13-branched, to 1.6 mm long; veins 17–191, unequal. Inflorescences 4–9, 11–30 cm long and 2–4 mm wide. Flowers (2–)4–14 in a raceme; pedicels 3–23 mm long; bracts subulate, acute, keeled, 0.9–2.5 cm long and 1.5–4 mm wide, the lowest with 3 or 5 veins; tepals oblong or ovate, acute or the inner obtuse at apex, 12–15 mm long and 5–8 mm wide; stamens equal, 6–8 mm long, filaments subulate, 2.5–5 mm long, anthers 4.5–7 mm long, fused or slightly emarginate at apex; ovary 4–10 mm long and 3–5 mm wide, style three-cornered, 1–3 mm long, stigma pyramidal, obtuse, 2 mm long, composed of three linear papillose receptive surfaces or 3-lobed with unequal triangular lobes. Capsule obconical or turbinate, 4–10 mm long and 4–8 mm wide, circumscissile. Seeds ovoid or spherical, 1–1.5 mm long; testa bristly with pointed pyramidal projections, winged with cuticula (as in Fig. 1: 3 & 6, p. 4).

TANZANIA. Ufipa District: Chapota, 4 Dec. 1949, *Bullock* 2045! & Sumbawanga, 27 Nov. 1954, *Richards* 3444 pro parte!
DISTR. **T** 4; Congo-Kinshasa, Zambia, Malawi
HAB. Montane grassland, upland pasture; 2150–2300 m

SYN. *Hypoxis* sp. A of Nordal et al. in Nord. J. Bot. 5: 29 (1985) pro parte quoad *Bullock* 2045
    H. multiceps sensu Zimudzi in Kirkia 16: 14 (1997) quoad *Pawek* 4142, *non* Baker (1896)

NOTE. This species has two distinctive varieties, one with and one without indumentum on the upper lamina surface. Only the first variety was found in Tanzania. Formal taxonomic recognition is submitted for publication.

8. **Hypoxis polystachya** *Baker* in Trans. Linn. Soc., Bot 1: 266 (1878); Baker in J.L.S. 17: 115 (1878) & in F.T.A. 7: 382 (1898). Type: Angola, Huilla, *Welwitsch* 4060 (BM!, holo.)

Robust herb up to 100 cm tall. Corms subglobose to elongated, 5–11 × 4–5 cm, white or yellow inside. Leaves tristichous, arching when mature, lanceolate to broadly linear, 40–100 × 2–5.8 cm, lamina white pilose-pubescent abaxially (outside), becoming glabrescent with age, glabrous adaxially (sometimes with a few hairs), margins and abaxial midrib densely white-lanate; hairs silvery-white, up to ± 3 mm long, tufted, 3–10-armed with arms subequal ascending-spreading; veins 31–41, closely spaced. Inflorescences 1–7, scape up to 18 cm long, width reducing in steps as pedicels diverge. Flowers 8–26, in a dense cylindrical raceme, flowering sequence often undetermined; pedicels 2–10 mm long; bracts up to 20–30 mm long, 4 mm wide, subulate-lanceolate, exceeding the pedicels; tepals narrowly ovate-lanceolate, 12–17 × 5–7(–8) mm, the inner tepals wider than the outer; stamens equal, filaments subulate, 2.2–3 mm long, anthers 4.5–7 mm long with thecae fused; ovary obconical, 7 × 4 mm, style ± 2–6 mm long, stigma 1.5–4 mm long, narrowly pyramidal. Capsule turbinate-obconical, 6–8 × 3.5–5 mm, circumscissile. Seeds brownish, 1.6–1.8 mm diameter; testa papillose, most often each papilla with a minutely wrinkled cuticle and 3–4 longitudinal narrow wing-like ribs or with wrinkled cuticula without wings (as in Fig. 1: 9, and as in Fig. 1: 3 & 9, p. 4).

TANZANIA. Mbeya District: Mbozi, 1935, *Horsbrugh-Porter*! & 19 Nov. 1930, *Davies* 742! & Mshewe Village, 1 Mar. 2001, *Wiland & Mboya* 154!
DISTR. **T** 7; Congo-Kinshasa, Angola, Malawi
HAB. Miombo woodland; 1150–1800 m

SYN. *Hypoxis subspicata* Pax in E.J. 15: 143 (1892); Baker in F.T.A. 7: 381 (1898); Nel in E.J. 51: 320 (1914); Geerinck in F.A.C.: 6 (1971); Nordal & Zimudzi in F.Z. 12 (3): 7 (2001);

Wiland-Szymańska in Ann. Missouri Bot. Gard. 88: 341, figs. 10A, B (2001). Types: Angola, Quango, *Pogge* 424 (B!, syn.) & Malandsche, *Teuscz* in *von Mechow* 249 (B!, syn.) *Hypoxis* sp. A of Nordal et al. in Nord. J. Bot. 5: 29 (1985) pro parte quoad *Davies* 742

NOTE. The name *H. polystachya* was used for a different taxon in F.Z. (Nordal & Zimudzi 2001), i.e. for what here is referred to *H. fischeri* Pax, a taxon without cuticular folding on the seed testa. A thorough investigation of the type specimen of *H. polystachya* (*Welwitsch* 4060) revealed a kind of cuticular folding, although not the typical one with wing-like ribs.

9. **Hypoxis galpinii** *Baker* in F.C. 6: 188 (1896), as *galpini*; Nel in E.J. 51: 320 (1914); Zimudzi in Kirkia 16: 16 (1996); Nordal & Zimudzi in F.Z. 12, 3: 11 (2001). Type: South Africa, 'Transvaal', *Galpin* 1098 (K!, holo.; PRE!, iso.)

Medium robust herbs up to 40 cm tall. Corm subglobose, 4–5.5 × 1.5–3.7 cm, orange inside, crowned by black fibrous remnants of former leaves. Leaves dark green, drying red-brown, with sheathing basal part, forming a short pseudostem, erect, coriaceaous, lanceolate, ± flat, eventually conduplicate, 14–40 × 1–2.2 cm, with white hairs on margins and midrib abiaxially, otherwise glabrous on both surfaces or with sparse indumentum also on the lamina; hairs tufted, 3–9(–12)-armed, up to 5 mm long; veins 13–45, unequal. Inflorescences 2–7, with scapes 11.5–30 cm long, 1.5–3 mm wide. Flowers 4–10, in a spiciform arrangement; pedicels 2–8 mm long (upper flowers usually sessile); bracts 10–20 mm long, subulate to lanceolate; tepals 7–11 × 3–5(–6.5) mm, broadly elliptic, inner slightly wider than the outer; stamens unequal, outer 3.5–5.2 with filaments 1.7–2.2 mm long, inner 3.5–5 mm long with filaments 1.4–1.7 mm long, filaments subulate, anthers 2.5–4.8 mm long with thecae fused; ovary obconical 3–4 × 2–3 mm, style 0.8–2.5 mm long, stigma 1.2–2.5 mm long with 3 free or fused erect lobes. Capsule cylindric-turbinate to clavate, 5–10 × 4.5–5 mm, circumscissile. Seeds black and glossy, 1.5–1.6 × 1.2–1.4 mm, ovoid; testa most often papillate (as in Fig. 1: 2 & 5, p. 4), cuticle smooth.

TANZANIA. Morogoro District: Uluguru, Lukwangule Plateau, over Chenzema Mission, 13 Mar. 1953, *Drummond & Hemsley* 1565!; Iringa District: Ifiga, 11 Mar. 1934, *Troll* 5383!; Mbeya District: Usafwa, Ntumbi, 21 Dec. 1978, *Leedal* 5248!
DISTR. T 6, 7; Zimbabwe, South Africa
HAB. Upland grasslands; 1900–2600 m

SYN. *H. infausta* Nel in E.J. 51: 319–320 (1914). Type: Tanzania, Morogoro District, Uluguru, *Stuhlmann* 9161 (B!, holo.)

NOTE. Plants from East Africa are more slender and have narrower leaves than those of the type specimen from South Africa. The specimens from Iringa and Mbeya deviate by having hairs also on the lamina surface. Specimens similar to the Iringa collections are, however, also found on the Inyanga Mountains in Zimbabwe. So far we do not recommend taxonomic delimitation on the subspecific level.

10. **Hypoxis obtusa** *Ker Gawl.* in Bot. Reg.: tab. 159 (1816); Baker in J.L.S. 17: 114 (1878) pro parte quoad type & in F.T.A. 7: 382 (1898); Nel in E.J. 51: 334 (1914); Norlindh & Weimarck in Bot. Not. 1937: 167 (1937) pro parte; Nordal & Zimudzi in F.Z. 12, 3: 13 (2001). Type: South Africa, Bot. Reg. Tab. 159 (Iconotype!)

Moderately robust herb 16–50 cm tall. Corm ovoid or cylindrical, yellow-orange inside, 4–11 × 3.5–6 cm, crowned by black or brown fibrous remnants of former leaves, often very copious. Leaves forming a short pseudostem, tristichous, erect at first soon becoming arcuate to ± straggling, conduplicate at least towards the base, linear, 10–70 × 0.6–3 cm, glabrous on both surfaces or sometimes thinly loosely strigose-pubescent in young leaves, the margins and abaxial midrib with densely matted white strigose appressed tufted hairs, 3–6-armed; veins 33–70, uniformly broad and closely spaced. Inflorescences 1–13, usually overtopping the arcuate leaves, first appearing before the leaves, with scapes 13–30 cm long, 3–5 mm wide. Flowers 7–13, in a raceme; pedicels 4–23 mm long, decreasing in length upwards; bracts subulate to linear-lanceolate, 7–25(–30) mm long, ± 2 mm wide; tepals ovate, acute or obtuse, (7–)11–18(–22) × 3–10 mm, inner broader than the outer; stamens subequal or inner shorter than outer, outer 5–9 mm long with filaments 3.5–5 mm long, inner 4.5–7.5 with subulate filaments 2–3.5 mm long, anthers 3.5–8 mm long, thecae fused or slightly emarginate; ovary 5-7 mm × 4 mm, style 1–2 mm long, stigma 1.5–3 mm long. Capsule obconical, 6–10 × 4–7 mm, circumscissile. Seeds black and glossy, ovoid, 1.3–2 × 1–1.5 mm; testa covered in closely-spaced dome-shaped papillae (similar to Fig 1: 2 & 5, p. 4), cuticle smooth.

UGANDA. Acholi District: Lomwaga Mountain, 5 Apr. 1945, *Greenway & Hummel* 7279!
KENYA. Kapsiliat [not traced], 22 Dec. 1943, *Starzeński* s.n.!; Kipkarren District, Mar. 1932, *Brodhurst-Hill* 745!
TANZANIA. Masai District: Ol Doinyo Lolkisale, Feb. 1967, *Beesley* 230!; Njombe District: Lihogosa Swamp near Njombe, 18 Jan. 1957, *Richards* 7892!; Iringa District: Dabaga highlands, Kilolo, 38 km SE of Iringa, 8 Feb. 1962, *Polhill & Paulo* 1394!
DISTR. **U** 1; **K** 3; **T** 1–3, 6, 7; Congo-Kinshasa, Angola, Zimbabwe, Botswana, South Africa
HAB. Grassland, scattered tree grassland, swamp edge; 1700–2750 m

SYN. *H. angolensis* Baker in Trans. Linn. Soc. ser. 2, Bot. 1: 266 (1878); Baker in F.T.A. 7: 380 (1898); Wiland-Szymańska in Ann. Missouri Bot. Gard. 88: 306–309, figs. 1, 2 A &B (2001). Type: Angola, Huilla, near Lopollo, *Welwitsch* 4059 (BM!, holo.)
    H. demissa Nel in E.J. 51: 328 (1914) Type: Tanzania, Morogoro District, Uluguru, Uhehe, Usangu, *Prittwitz & Gaffron* 171 (B!, holo.)
    H. protrusa Nel in E.J. 51: 336 (1914). Types: Tanzania, Kilimanjaro, Umbugwe and Iraku, *Merker* 18 (B!, syn.) & Mbulu District, Wanege highlands, between Akida Maussa and Mangati, *Jaeger* 243 (B!, syn.)
    H. obtusa complex sensu Nordal et al. in Nordic J. Bot. 5: 28 (1985) pro parte
    H. villosa complex of Zimudzi in Kirkia 16: 14 (1996) pro parte

11. **Hypoxis fischeri** *Pax* in E.J. 15: 143 (1893); Baker in F.T.A. 7: 382 (1898); Nordal & Iversen in Fl. Cameroun 30: 34, t. 8 & 13C (1987). Type: East Africa, probably lake region [Seengebiet], 26–30 Mar., *Fischer* 611 (B!, holo.; K!, iso.)

Robust or medium herb up to 78 cm tall. Corm subglobose to cylindrical, 3–9 × 3–6 cm, white or yellowish inside, crowned with fibrous remnants of former leaves. Leaves tristichous, forming a short pseudostem, ± rigidly erect, becoming lax, usually conduplicate, linear to lanceolate, 10–78(–102) × 0.8–5.7 cm, always ciliate on margin & midrib with tufted hairs, on the lamina pubescent on the both surfaces or only below, with tufted to 8-armed or bifurcate silvery-strigose hairs; veins unequal, 11–40(–70). Inflorescences 1–10(–12), with scapes up to 37(–50) cm long, overtopped by the mature leaves, 2–5 mm wide, with reducing in steps as pedicels diverge. Flowers 3–14(–25) in a raceme-like ± cylindrical arrangement, floral anthesis acropetal; pedicels 2–25 mm long, decreasing in length upwards; bracts 7–2.7 mm long, subulate to linear-lanceolate; tepals (9–)10–18 × (3–)4–8 mm, inner slightly broader than the outer; stamens equal 6–7 mm long, filaments subulate to almost linear, 1.3–4 mm long, anthers 3–8 mm long, entire at apex; ovary 3–9 mm × 3–5 mm, style 1–6 mm long, stigma 1–4 mm long, composed of 3 linear lobes fused to apex or 3 free triangular lobes. Capsule turbinate, 4–9 × 3–6 mm with opening by transverse slits. Seeds black and most often glossy, (sub-)globose 1.2–2 mm diameter, testa variable (see key below), but never with cuticular folding.

NOTE. This taxon is heterogenous, particularly as to leaf indumentum and testa structure. Distinct forms can be recognised (probably apomictic forms) and are here referred to the rank of variety. Unfortunately specimens lacking seeds can be only referred to as *H. fischeri* sensu lato. Such is the case for specimens from **K** 3, 4/6; **T** 4, 7 – none of which are referrable to variety.
    Though the protologue says Fischer 60, not 611, the number on the type specimens, which bear the correct date, is 611

Distribution for *H. fischeri* sensu lato: **K** 3, 4/6; **T** 4, 7; Cameroon, Congo-Kinshasa, Zambia, Malawi, Mozambique

This species was unfortunately referred to *Hypoxis polystachya* in F.Z. (Nordal & Zimudzi 2001) due to misinterpretation of testa structure. This lacks cuticular folding in *H. fischeri*, as opposed to *H. polystachya*.

1. Seed testa flat and dull; all hairs on the lamina tufted ... b. var. *colliculata*
   Seed testa distinctively papillate, glossy; trichomes on the lamina tufted or two-branched except on edges and midrib 2
2. Seed papillae acute at apex; trichomes on the lamina 2–3-branched except on edges and midrib 3
   Seed papillae obtuse at apex; trichomes on the lamina tufted or two-branched except edges and midrib 4
3. Leaves 8–14 mm wide, sparsely hairy; inflorescence 10–13-flowered; pedicels 7–15 mm long; papillae conical, long-acuminate . . . . c. var. *katangensis*
   Leaves about 30 mm wide, hairy mainly on the lower side of the blade; inflorescence 12–30-flowered; pedicels 3–10 mm long; papillae mammiform . . . . a. var. *fischeri*
4. Leaves 8–15 mm wide, usually densely villous with white long tufted trichomes; inflorescence 9–11-flowered . . . . d. var. *hockii*
   Leaves 12–20 mm wide, pubescent, trichomes white or yellowish, two-branched except on edges and midrib; inflorescence 6–7-flowered . . . . e. var. *zernyi*

a. var. **fischeri**

Corm deep ochre; robust plants with leaf width about 3 cm, bifurcate hairs on lamina on both sides of the leaf; papillate seeds with mammiliform seeds papillae (Fig. 1: 8, p. 4).

UGANDA. Acholi District: 2 km NE of Lotututru, 17 Feb. 1969, *Lye & Lester* 2079!; Karamoja District: Namojongotyang, Mt Kadam [Debasien], no date, *Eggeling* 2815!; Mbale District: Mt Elgon, Sipi, 17 Feb. 1924, *Snowden* 822!
KENYA. Trans-Nzoia District: Kitale, 5 Mar. 1955, *Bogdan* 3691! & Apr. 1962, *Tweedie* 2341! & *Tweedie* 2343!
TANZANIA. Biharamulo District: Kashasha, 10 Aug. 1960, *Tanner* 5071!; Ufipa District: Mbisi, 6 Oct. 1950, *Bullock* 3413!; Mbeya District: Mbozi, 14 Nov. 1932, *Davies* 753!
DISTR. **U** 1, 3; **K** 3; **T** 1, 4, 7; Cameroon
HAB. Grassland, wooded or scattered tree grassland, recently burnt terrain; 1500–2600 m

SYN. *H. multiflora* Nel in E.J. 51: 317 (1914); F.P.S. 3: 306 (1956). Type: Uganda, 'Elgon-District', *Evan James* s.n. (K!, holo.)
 H. obtusa complex sensu Nordal et al. in Nordic J. Bot. 5: 28 (1985) pro parte
 H. polystachya sensu Nordal & Zimudzi in F.Z. 12, 3: 14–15 (2001) pro parte

b. var. **colliculata** (*Wiland*) Wiland & Nordal **comb. nov.** Type: Congo-Kinshasa, Shaba, Kundelungu Plateau, R. Lofoi, *Lisowski, Malaisse & Symoens* 7682 (POZG!, holo.)

Flesh of the corm greenish cream. Leaves linear to lanceolate, 0.8–2 cm wide, tufted hairs along margins and on lamina. Seeds dull, testa without distinct papillae (similar to Fig. 1: 7, p. 4, but with micropapillation).

TANZANIA. Ufipa District: Sumbawanga, 28 Nov. 1954, *Richards* 2379A! & Mbisi, 6 Oct. 1950, *Bullock* 3414!; Iringa District: N of escarpment forest & S of Iringa–Mbeya Road between Nyololo and Mafinga [James & John's Corner], 29 Dec. 1973, *Spjut & Muchai* 3467!
DISTR. **T** 4, 7; Congo-Kinshasa
HAB. Upland pasture, on fireswept eroded hillside, among rocks; 2300–2500 m

SYN. *H. hockii* De Wild. var. *colliculata* Wiland in Ann. Missouri Bot. Gard. 88: 321–324, figs. 9 A–E, 10 C & D (2001)

c. var. **katangensis** (*De Wild.*) *Wiland & Nordal* **comb. et stat. nov**. Types: Congo-Kinshasa, Shaba [Katanga], *Verdick* s.n. (BR!, syn.); Zambia, between Lakes Bangweulu [Banguelo] and Tanganyika, *Fries* 1148 (UPS!, syn.)

Leaves linear, 0.8–1.4 cm wide, prominently pubescent on margins and the midrib below with tufteded hairs, on surface only sparsely pubescent with 2- or 3-branched hairs. Seeds glossy, testa with long-acuminate conical papillae (similar to Fig. 1: 8, p. 4, but papillae not so round).

Tanzania. Ufipa District: Msambia [Msamvia], 15 Nov. 1908, *Münzner* 59!
Distr. **T** 4; Congo-Kinshasa, Zambia
Hab. Grassland; 1800 m

Syn. *H. katangensis* De Wild. in E.J. 51: 312 (1914)
   H. aculeata Nel in E.J. 51: 327 (1914). Type: Tanzania, Ufipa District, Msamvia, *Münzner* 59 (B!, holo.)
   H. obtusa complex sensu Nordal et al. in Nordic J. Bot. 5: 28 (1985) pro parte
   H. polystachya sensu Nordal & Zimudzi in F.Z. 12, 3: 14–15 (2001) pro parte
   H. hockii De Wild. var. *katangensis* (De Wild.) Wiland in Ann. Missouri Bot. Gard. 88: 324, figs. 9G–J, 10E (2001)

d. var. **hockii** (*De Wild.*) *Wiland & Nordal* **comb. et stat. nov**. Type: Congo-Kinshasa, upper Shaba [Katanga], *Hock* s.n. (BR!, holo.; B!, iso.)

Corm white, cream, or yellow inside. Leaves light green, 0.8–1.5 cm wide, tufted hairs along margin, midrib and on leaf surfaces. Seeds glossy, testa papillate (as in Fig. 1: 2 & 5, p. 4).

Uganda. West Nile District: Koboko, without date, *Eggeling* 1840 pro parte! & Mar. 1938, *Hazel* 441!; Busoga District: 1 mile E of Nankoma Hill, 4 miles SSE of Kyemure, 16 Apr. 1953, *Wood* 992!
Tanzania. Ufipa District: Chala Mt, 10 Dec. 1956, *Richards* 7207!; Iringa District: Mufindi, N of escarpment forest & S of Iringa-Mbeya Road between James & John's Corner, vicinity of Ngwazi Estate, 29 Oct. 1973, *Spjut & Muchai* 3460!; Songea District: Myangayanga Mountains, above Myangayanga village, 21 Feb. 2001, *Wiland & Mboya* 105!
Distr. **U** 1, 3; **T** 4–8; Congo-Kinshasa
Hab. Grassland with scattered shrubs, miombo woodland; 1400–2300 m

Syn. *H. hockii* De Wild. in F.R. 11: 537 (1913); Wiland-Szymańska in Ann. Missouri Bot. Gard. 88: 320 (2001)
   H. pedicellata Nel in E.J. 51: 315 (1914). Types: Congo-Kinshasa, Lualuba, *Deschamps* s.n. (BR!, syn.) & Zambia, Mbala District, Msisi [Sisya] Forest, *Fries* 1261 & 1261a (UPS!, syn.)
   H. polystachya sensu Nordal & Zimudzi in F.Z. 12, 3: 14–15 (2001) pro parte

e. var. **zernyi** (*Schulze*) *Wiland & Nordal* **comb. et stat. nov**. Type: Tanzania, WSW of Songea, Mt Lupembe, *Zerny* 20 (W!, holo.; B!, iso.)

Plants growing separately or in clumps. Corm yellow or orange inside, getting bluish exposed to air, with a light acrid smell. Leaves yellow-green, reddish in basal part, 2.4–5.7 cm wide, tomentose, whitish or yellowish bifurcate hairs on lamina. Seeds glossy, testa finely papillose (as in Fig. 1: 2 & 5, p. 4). Fig. 2, p. 15.

Tanzania. Ufipa District: Mt Malonje, escarpment near Sumbawanga, 1 Jan. 1967, *Schultze* 238!; Iringa District: N of escarpment forest & S of Iringa–Mbeya Road, between Nyololo and Mafinga [James & John's Corner], 29 Oct. 1973, *Spjut & Muchai* 3461!; Mbeya District: Pungaluma Hills above Mshewe Village, 25 Nov. 1989, *Lovett & Kayombo* 3473!
Distr. **T** 2, 4, 7, 8; Zambia, Malawi, Mozambique
Hab. Miombo woodland, grassland; 1400–2600 m

Syn. *H. zernyi* Schulze in N.B.G.B. 14: 375 (1939)
   H. matangensis Schulze in N.B.G.B. 14: 376 (1939). Type: Tanzania, Songea District, WSW of Songea, Linda–Uyangayanga road, *Zerny* 374 (B!, holo.)
   H. obtusa complex sensu Nordal et al. in Nordic J. Bot. 5: 28 (1985) pro parte

Fig. 2. *HYPOXIS FISCHERI VAR. ZERNYI* — **1**, habit; **2**, trichome from scape; **3**, trichome from leaf edge; **4**, trichome from leaf lamina. 1 from *Lovett & Kayombo* 3420; 2–4 from *Wiland & Mboya* 132. Drawn by Kinga Gawrońska.

12. **Hypoxis gregoriana** *Rendle* in J.L.S. 30: 408 (1895); Baker in F.T.A. 7: 382 (1898). Type: Kenya, Naivasha District, Kikuyu Escarpment, Kedong, *Gregory* s.n. (BM!, holo.)

Medium robust plants up to 60 cm. Corm elongated, 8.5–10 × 1–3.4 cm, bright yellow or orange inside, crowned with fibrous black remnants of old leaf bases. Leaves creating a short pseudostem, linear, 10–75 × 1–2.5 cm, recurved, covered with white appressed or spreading tuft trichomes, mostly on abaxial side; hairs 3–5 branched, 1.5–4 mm long; veins 9–11, equal, spaced. Inflorescences 1–6, 5–55 cm long and 1–2 mm wide. Flowers 2–4(–6) in a corymbose raceme 3–4 cm long; pedicels 2–3 cm long; bracts subulate, 7–8 × 1.5 mm wide at base; tepals linear or lanceolate, acute, 8–14 × 2–4 mm; stamens unequal, outer stamens 4–4.3 mm long with filaments ± 3 mm long, inner stamens 3 mm long with subulate filaments 2 mm long, anthers 3–7 mm long, emarginate at apex; ovary obconical, 3–4 mm long, style cylindrical ± 0.5 mm long, stigma conical, thick, 1 mm long. Capsule obovoid, 5–7 mm long, 4–5 mm wide; seeds black, seed coat flat without distinct papillae, but with micropapillation (as in Fig. 1: 1 & 4, p. 4)

UGANDA. Mengo District: near Bukomero, Singo, 85 km Kampala–Hoima Road, Sep. 1932, *Eggeling* 931!; Masaka District: Mawogola Country, 17–18 km SE of Ntusi, 19 Oct. 1969, *Lye & Rwaburindore* 4517!
KENYA. Northern Frontier District: Mt Kulal, Apr. 1959, *Adamson* K5B!; Nairobi District: Bahati, May 1933, *Rogers* 479!; Masai District: between Namanga and Kajiado, 105 km on Nairobi–Namanga road, 17 Dec. 1961, *Polhill & Paulo* 1016!
TANZANIA. Musoma District: Kleins Camp to Wogakuria Hill, 30 Dec. 1964, *Greenway & Myles Turner* 12000!; Arusha District: Songe Hill, 23 Feb. 1969, *Richards* 24170!; Masai District: Ketumbeine Forest Reserve, N slope of Ketumbeine Mountain, S of Losiwira village, 18 Dec. 2001, *Gereau* 6583!
DISTR. **U** 4; **K** 1, 3, 4, 6; **T** 1, 2; not known from elsewhere
HAB. Grassland, either dry or swampy (e.g. black cotton soil), burnt grassland, wooded or scattered tree grassland; 1000–3650 m

SYN. *H. araneosa* Nel in E.J. 51: 310 (1914). Type: Tanzania, Masai District, Engare Nairobi [Ngare-Nairobi], *Endlich* 238 (B!, holo.)
    H. obtusa complex sensu Nordal et al. in Nordic J. Bot. 5: 28 (1985) pro parte

NOTE. This species closely resembles *H. urceolata* in morphology, but differs by having generally fewer flowers in a lax corymb; and in the seed surface structure (Fig. 1: 1 & 4, p. 4). In *H. gregoriana* all hairs covering the leaf lamina are tufted, whereas in *H. urceolata* this type of trichome occurs only on the midrib and edges of a leaf.

13. **Hypoxis nyasica** *Baker* in K.B. 1897: 284 (1897) & in F.T.A. 7: 380 (1898); Nel in E.J. 51: 318 (1914); Brenan in Mem. N.Y. Bot. Gard. 9: 86 (1954); Nordal & Zimudzi in F.Z. 12 (3): 16 (2001). Type: Malawi, Mount Zomba, *Whyte* s.n. (K!, B!, syn.) & Mount Malosa, *Whyte* s.n. (K!, syn.)

Slender herb, 15–67 cm tall, occasionally tufted with several pseudostems on a rootstock. Corm 1.5–3(–11) × 1–2.5(–5) cm, light yellow to deep orange inside. Leaves linear, grass-like, erect or sometimes straggling, conduplicate at least towards the base, to 67 × 0.3–0.8(–2) cm with length depending on stage, thinly pubescent on margins and main veins with 2–3-branched golden brown to whitish hairs, lamina glabrescent, sometimes with a few bifurcate hairs; veins 11–13, distinct, different in size and with variable distance between them. Inflorescences 1–8, overtopped early by the leaves, with scapes up to 42 cm long, 1–2(–3) mm wide, width reducing in steps as pedicels diverge. Flowers (2–)4–5(–10) in a raceme-like arrangement 2–9 cm long; pedicels 3–10(–20) mm long; bracts subulate to linear-lanceolate, 7–30 mm long; tepals all similar, elliptic to lanceolate, 8–11 × 3–6 mm; stamens equal, filaments

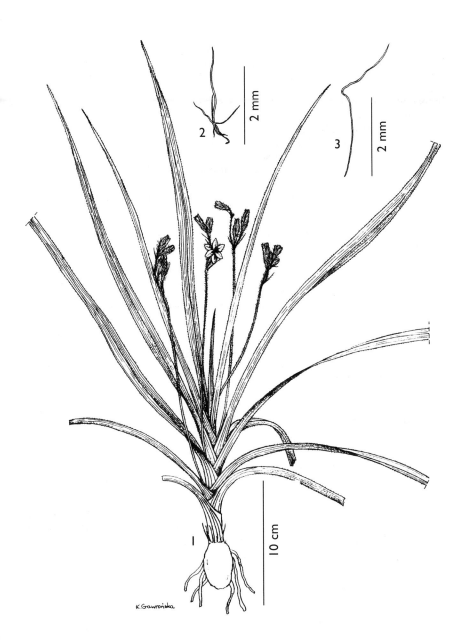

FIG. 3. *HYPOXIS NYASICA* — **1**, habit; **2**, trichome from scape; **3**, trichome from leaf edge. 1 from *Holz* 537; 2–3 from *Richards* 11701. Drawn by Kinga Gawrońska.

subulate, 1.5–3 mm long, anthers 2.3–5 mm long, emarginate at apex; style columnar, 1–3 mm, stigma pyramidal, 1–2 mm long, with three rows of papillae. Capsule 4–6(–8) × 3–4.5 mm, turbinate, opening with a transverse slit near the middle. Seeds black and glossy, 1–1.2 × 1.7 mm, subglobose; testa densely papillate, the papillae dome-shaped, the cuticle smooth. (Fig. 1: 2 & 5, p. 4). Fig. 3, p. 17.

TANZANIA. Uzaramo District: East of coast range, Nov. 1860, *Speke & Grant* s.n.!; Iringa District: Mufindi, Kigogo, 9 Mar. 1987, *Lovett, Keeley & Niblett* 1672!; Lindi District, Nachingwea, 21 Sep. 1952, *Anderson* 792!
DISTR. **T** 4, 6, 7, 8; Zambia, Malawi, Mozambique, Zimbabwe
HAB. Grassland, *Brachystegia* woodland, open woodland or disturbed forest; 400–2300 m

SYN. *H. campanulata* Nel in E.J. 51: 314 (1914). Type: Tanzania, Lindi District, Tendaguru, *Janensch & Hennig* s.n. (B!, holo.)
    H. engleriana Nel in E.J. 51: 315 (1914). Types: Malawi, Blantyre, Shire Highlands, *Buchanan* 26 (K!, syn.) & Shire Highlands, Zambesiland, *Geo Adamson* 28 (K! syn.) & Blantyre, *Scott* s.n. (K! syn.)
    H. engleriana Nel var. *scottii* Nel in E.J. 51: 315 (1914). Type: Malawi, Shire Highlands, *Scott-Elliot* 8579 (K!, holo.)
    H. ingrata Nel in E.J. 51: 311 (1914). Type: Malawi, lower plateau, north of Lake Malawi (Nyassa), *Rev. Thomson* s.n. (K!, holo.)
    H. probata Nel in E.J. 51: 317 (1914). Type: Tanzania, Rungwe District, Kyimbila, *Stolz* 537 (B!, holo.; G!, K!, L!, M!, S!, iso.)
    H. canaliculata sensu Brenan in Mem. N.Y. Bot. Gard. 9: 86 (1954) quoad *Brass* 17598, *non* Baker
    H. urceolata sensu Cribb et Leedal in Mount. Fl. S. Tanz.: 167, Pl. 46C (1982), *non* Nel
    H. obtusa complex sensu Nordal et al. in Nordic J. Bot. 5: 28 (1985) pro parte
    H. villosa sensu Zimudzi in Kirkia 16: 17 (1997) pro parte; sensu Rendle in J. L. S., Bot. 40: 211 (1911); sensu Eyles in Trans. Roy. Soc. S. Afr. 5: 328 (1916) pro parte quoad *Rogers* 4043

NOTE. This species is closely related to *H. urceolata*, but has generally narrower leaves and more distinctly papillose seeds. Young stages of *H. urceolata* may easily be confused with mature *H. nyasica*. The taxon, as defined here, is vicariant with *H. urceolata* occurring north of *H. nyasica*.

14. **Hypoxis urceolata** *Nel* in E.J. 51: 336 (1914); Hutchinson & Dalziel in F.W.T.A. 2: 394 (1936); Jex-Blake in W.F.K. 130 (1948); F.P.S. 3: 306 (1956); Morton in W. Afr. Lilies and Orchids: 31, t. IX, fig. 35 (1968); Hepper in F.W.T.A. ed. 2, 3: 172 (1968); Troupin in Syll. Fl. Rwanda: VI.277, fig. p. VI.276 (1971); Wiland-Szymańska in Ann. Missouri Bot. Gard. 88: 345, figs. 18A–G, 19 (2001). Types: Tanzania, sine loc., *Obst* s.n. (B!, syn.); Uganda: Ruwenzori, Buddu, *Dawe* 231 (K!, syn.) & Ohagwe, *Dawe* 103 (K!, syn.); Kenya: Nairobi, *Whyte* s.n. (K!, syn.)

Robust herb up to 60 cm tall. Corm ovoid, 3.3–6.5 × 2.3–3(–4.3) cm, yellow to orange (rarely whitish) inside, crowned by fibrous leaf remnants. Leaves linear to narrowly lanceolate, (11–)17–60(–95) × (0.8–)1.2–2 cm, often reflexed; thinly pubescent with tufted golden brown to whitish hairs on the margins and main veins, lamina glabrescent, sometimes with a few bifurcate hairs present; veins unequal, (12–)19–45. Inflorescences 4–9, with scapes 8–45 cm × 2–3 mm. Flowers (2–)4–7 in a racemose arrangement; pedicels 0–35 mm long; bracts subulate, keeled, (5–)9–24 × 1–2 mm; tepals ovate, 8–15(–17) × 4–5(–6) mm; stamens unequal, outer stamens 4–8 mm, inner 4–7 mm long, filaments subulate, 1.5–4 mm long, anthers 3–6 mm long, fused apically; ovary obconical, 4–10 × 3–4 mm long, style 3–4 mm long, stigma pyramidal, 2–3 mm long, with three lines of papillae. Capsule turbinate, 4–10 × 3–5 mm, opening with a transverse slit. Seeds black, ovoid, ± 1.5 × 1 mm; seed coat without distinct papillae. (Fig. 1: 7, p. 4)

UGANDA. Ankole District: Ruizi River, 1 Dec. 1950, *Jarrett* 144!; Toro District: Central Kibale, 10 Oct. 1940, *Sangster* 683!; West Nile District: Lendu, Apr. 1940, *Eggeling* 3893!
KENYA. Nairobi, 29 Jan. 1932, *Napier* 1777!; Machakos District: Chyulu north, 19 Apr. 1938, *Bally* 735!; Masai District: Mara Game Reserve, 15 July 1968, *Agnew & Braun* 10187!

TANZANIA. Bukoba District: Bunazi, Oct. 1931, *Haarer* 2334!; Lushoto District: Korogwe – Handeni road, 22 Apr. 1954, *Faulkner* 1420!; Mpanda District: Ikola–Mpanda road 34 km from Ikola, 8 Nov. 1959, *Richards* 11750!

DISTR. **U** 1–4; **K** 1–7; **T** 1–6; Congo-Kinshasa, Rwanda

HAB. Grassland, disturbed bushland, open woodland; 1300–3100 m

SYN. *H. apiculata* Nel in E.J. 51: 327 (1914); Cufodontis in E.P.A. 41: 1577 (1971). Type: Kenya, Teita District: N'di, *Hildebrandt* 2542 (B!, holo.)
   H. arenosa Nel in E.J. 51: 310 (1914). Type: Tanzania, Lushoto District, E Usambara, *Holst* 93 (B!, holo)
   H. crispa Nel in E.J. 51: 334 (1914). Type: Tanzania, Kilimanjaro, Muengue [Ngowe] area, *Volkens* 360 (B!, holo.)
   H. cryptophylla Nel in E.J. 51: 316 (1914). Type: Tanzania, Bukoba District: S Karagwe, W bank of Lake Victoria [Nyansa] to Kagera, *von Trotha* 140 (B!, holo.)
   H. textilis Nel in E.J. 51: 326 (1914). Type: Tanzania, Mwanza District, Bumpeke [Umpeke] *Stuhlmann* 858 a (B!, holo.)
   H. bequaertii De Wild. in Pl. Bequaert. 1: 49 (1921); Robyns & Tournay in F.P.N.A. 3: 390 (1955). Types: Congo-Kinshasa, upper Congo, between Irumu and Bogoro, *Bequaert* 4919 (BR!, syn.) & Kivu, between Beni and Kasindi, *Bequaert* 5198 (BR!, syn.)
   H. obtusa complex sensu Nordal et al. in Nordic J. Bot. 5: 28 (1985) pro parte
   H. obtusa sensu Champluvier in Fl. Rwanda 4: 84 (1987), *non* Ker-Gawl.

NOTE. See *H. nyasica* and *H. gregoriana.*

15. **Hypoxis rigidula** *Baker* in J.L.S. 17: 116 (1878) & in F.C. 6: 186 (1896); Nel in E.J. 51: 331 (1914); Norlindh & Weimarck in Bot. Not. 1937: 166 (1937); Zimudzi in Kirkia 16: 16 (1996) pro parte; Nordal & Zimudzi in F.Z. 12, 3: 12 (2001). Types: South Africa, *Cooper* 883 (K, syn.) & 3239 (K, syn.) & 1763 (?, syn.) & 3241 (?, syn.); *Burchell* 3694 (K, syn.); *Zeyher* 1670 (K, P!, syn.); *Drège* 2194 (K, P!, syn.); *Hort. Kew* anno 1863 (K, syn.)

Robust herb to 1 m tall, growing separately or in clumps. Corm turbinate to hemispheric with flat base, vertical or horizontal, 2.5–10.5 × 1.5–6 cm, yellow or orange inside. Leaves dark green, lighter at base, creating a distinctive pseudostem 6–15 cm long, linear, rigidly coriaceous, erect, conduplicate, 30–130 × 0.5–2.2 cm, pilose with tufted whitish to silvery greyish hairs along margins and midrib and with 2–3-branched hairs on the lamina; veins 11–17, distinctive. Inflorescences 1–5, with scapes 30–70 cm × 1.5–2 mm, overtopped by the mature leaves. Flowers 3–14(–23), alternate, in an up to 15 cm long spicate raceme; lower pedicels 0.6–1(–1.9) cm long, shorter towards the apex; bracts linear-subulate, 12–26 mm long; tepals lanceolate, acute, (5–)12–18 × 5–7 mm; stamens equal, filaments subulate, 1.3–2.5 mm long, anthers 3–6 mm long, fused apically; ovary turbinate, 4–8 × 3–5 mm, style 0.5–1(–2.3) mm, stigma 1–2(–3.5) mm long. Capsule turbinate, 8 mm long, opening by transverse slit near the middle. Seeds glossy, black, globose, 1.2–1.7 mm diameter with papillate testa (as in Fig 1: 2 & 5, p. 4), cuticle smooth.

KENYA. West Suk District: Keringet, June 1969, *Tweedie* 3657!; Kisumu-Londiani District: Tinderet Forest Reserve, 26 June 1949, *Maas Geesteranus* 5193!; Taita District: Mbololo, June–July 1938, *Boy Joanna* in *CM* 9023!
TANZANIA. Lushoto District: West Usambara Mts, below Baga II Forest Reserve, between Mgwashi and Mtai above Mzinga village, 31 Jan. 1985, *Borhidi, Iversen & Mziray* 85425!; Mpwapwa District: Kiboriani Mountains, 9 Dec. 1938, *Hornby & Hornby* 2097!; Mufundi District: N of Escarpment Forest & S of Iringa–Mbeya road, vicinity of Ngwazi Estate, 29 Oct. 1973, *Spjut & Muchai* 3466!
DISTR. **K** 3–5, 7; **T** 1, 3, 5, 6, 7; Mozambique, Zimbabwe, Swaziland, Lesotho, South Africa
HAB. Open grassland or woodland, fallows; laterite or black soil; 800–2300 m

SYN. *H. laikipiensis* Rendle in J.L.S. 30: 407 (1895); Baker in F.T.A. 7: 381 (1898). Type: Kenya, Laikipia District, W of Alng'aria, *Gregory* s.n. (photo.!)
   H. obtusa complex sensu Nordal et al. in Nordic J. Bot. 5: 28 (1985) pro parte

2. CURCULIGO

Gaertn. in Fruct. Sem. Pl. 1: 63, tab. 16 (1788); Baker in J.L.S. 17: 122 (1878); Baker in F.T.A. 6: 382–383 (1898); Nel in E.J. 51: 258–259 (1914); Nordal in Kubitzki, Fam. Gen. Vasc. Pl. 3: 293 & Fig. 88E–I (1998); Wiland in Fragm. Fl. Geobot. 42: 9–24 (1997)

Rhizome elongated and vertical, often branched, with fleshy roots. Leaves ± pseudopetiolate with sheathing leaf bases; lamina plicate and sparsely pilose. Scapes short, most often completely enveloped by cataphylls. Flowers subsessile, most often single, sometimes few in a ± umbellate inflorescence, supported by leafy involucral bracts; perianth segments free, patent, yellow; filaments filiform or subulate, attached in the sinuses between anther theca; anthers sagittate with latrorse opening. Ovary trilocular, surrounded by bracts and old leaf remnants, often subterranean, separated from the perianth by a conspicuous beak. Fruit indehiscent, more or less succulent at maturity, crowned with the persistent ovary beak. Seeds with a conspicuous, swollen and hooked funiculus.

Curculigo pilosa (*Schum. & Thonn.*) *Engl.* in V.E. 2: 353 (1908); Hutchinson & Dalziel in F.W.T.A. 2: 396 pro parte & Fig. 319 (1936); Robyns & Tournay in Fl. Sperm. Parc. Nat. Albert 3: 387 (1955); Troupin in Fl. Sperm. Parc Nat. Garamba 4: 206–207 (1956); Hepper in F.W.T.A., 2 ed., 3: 174 pro parte (1963); Morton in W. Afr. Lilies and Orchids: 31–32 & Pl. X, 37 (1968); E.P.A.: 1577 (1971); Geerinck in F.C.B.: 2–4 & fig. p. 3. (1971); Wickens in Fl. Jebel Marra: 159 (1976); Zimudzi in Kirkia 16: 13 (1996); Nordal in Fl. Eth.: 89 & Fig. 189.2 pro parte (1997); Wiland in Fragm. Fl. Geobot. 42: 9–24 (1997). Type: Ghana, *Thonning* s.n. (L, holo.; K!, photo.)

Slender to medium robust herb, up to 0.5 m high. Rhizome fusiform or ovoid, cream inside, 1.4–11 cm long and 3–30 mm wide, sometimes divided in two; roots thick, ribbon-like; rhizome surmounted by old leaf-bases, for up to 12 cm. Leaves sometimes creating a short white pseudostem, often reddish at base, linear or lanceolate, 6–51(–75) cm long and (2–)4–20 mm wide, narrowing towards the apex, acute, sessile or subsessile, pilose on both sides or almost glabrous; hairs tufted, 6–10-branched; veins 5–65(–81). Scapes 1–5, 1.2–4.7(–9) cm long and 0.5–2.5 mm wide, pilose. Flowers yellow, orange or pinkish, solitary or rarely in 2–5-flowered racemes, hermaphrodite or male with a reduced pistil; hermaphrodite flowers usually subtended by two bracts, the upper sometimes absent; occasionally a third wide-lanceolate bract, enclosing the ovary is present; lower bract linear or lanceolate, acuminate, forming a fused spathe round the ovary 1.9–7.1 cm long and 3–10 mm wide, ciliate at least on the margins, 11–39-veined; upper bract filiform, linear or subulate, acute, 1.3–5.9 cm long and 0.2–3 mm wide, pilose, 7–13-veined; in inflorescences each flower is subtended by only one bract, never fused in the basal part, 0.8–2.7 cm long and 1.3–4 mm wide, about 5-veined. Tepals ovate or lanceolate, acute, (6–)7–20 mm long and 2–5 mm wide, outer pilose outside; stamens equal, 3–8 mm long, filaments subulate, 1.5–6 mm long, anthers oblong, sagittate, fused or emarginate at the apex, 2–4 mm long; ovary fusiform, 5–19 mm long, 1–2 mm wide, pilose or subglabrous, rostrate, rostrum 1.6–7.5 cm long, pilose, style columnar, 3–14.5 mm long, stigma of 3 fused or free oblong and obtuse lobes, 0.5–2 mm long. Fruit subterranean, indehiscent, fusiform, 9–33 mm long and 3–7 mm wide, pilose. Seeds ovoid, with a recurved and swollen funicle, 3–5 mm long, 2–3 mm wide, black and shining, sometimes striate.

SYN. *Gethyllis pilosa* Schum. & Thonn. in Beskr. Guin. Pl. : 192 (1828)
    Curculigo gallabatensis Baker in Trans. Linn. Soc., Bot. 1: 266 (1878); Baker in J.L.S. 17: 123 (1878) & in F.T.A. 6: 383 (1898); Durand & Schinz in Consp. Fl. Afr. 5: 236 (1895); Rendle in Hiern in Cat. Afr. Pl. Welw. 2: 31 (1899); De Wildeman in Ann. Mus. Congo, Bot., sèr. 5, 3: 352 (1912) & Pl. Bequaert. 1: 48 (1921); Nel in E.J. 51: 258 (1914:). Type: Sudan, Gallabat, Gendua, *Schweinfurth* 39 (K!, iso.; P!, BM!, iso.)

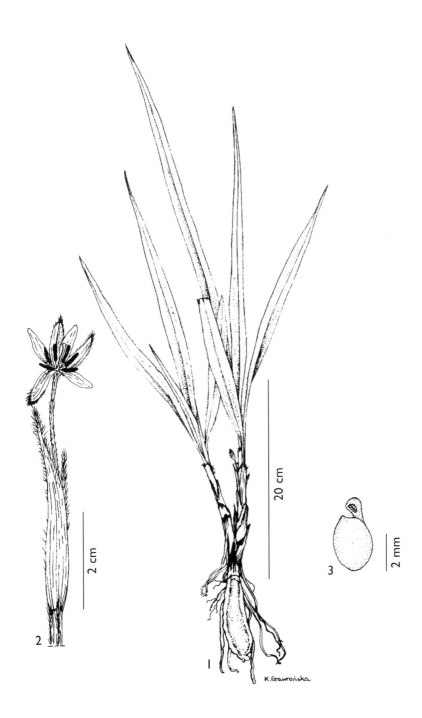

FIG. 4. *CURCULIGO PILOSA SUBSP. PILOSA* — **1**, habit; **2**, flower; **3**, seed. All from *Wiland & Mboya* 164. Drawn by Kinga Gawrońska.

1. Flowers 1 or in 2–5-flowered inflorescences; leaves linear,
   1–5 mm wide; rostrum 1.6–6 cm long; perianth segments
   7–10 mm long . . . .   b. subsp. *minor*
   Flowers 1 or 2, never in larger inflorescence; leaves linear or
   lanceolate, 3–20 mm wide; rostrum 2.4–7.5 cm long;
   perianth segments 10 mm long or longer 2
2. Leaves 4–20 mm wide, usually lanceolate; rostrum 2.4–2.9 cm
   long; perianth segments 10–12 mm long . . . .   a. subsp. *pilosa*
   Leaves 3–10 mm wide, linear; rostrum 5.5–7.5 cm long;
   perianth segments 17–20 mm long . . . .   c. subsp. *major*

a. subsp. **pilosa**

Plant up to 0.5 m high. Rhizome 2–9 cm long and 4–30 mm wide. Leaves (6.5–)11.5–51(–75) cm long and (2–)4–21 mm wide; veins (15–)18–65(–81). Scapes 1–4, usually 1-flowered, rarely 2-flowered, 1.5–4.3(–9) cm long and 1–2.5 mm wide. Flowers bracteate; lower bract 2.5–7.1 cm long and 5–10 mm wide, 13–39 veined; upper bract (2–)2.5–4.9 cm long, 1.5–3 mm wide, 7–9-veined; perianth segments (7–)10–12(–15) mm long and 2–3.5(–4) mm wide; stamens 5–6 mm long, filaments 2–3.5 mm long; anthers 2.5–4 mm long, emarginate at apex; ovary 9–13 mm long and 1.5–2 mm wide, rostrum 2.4–2.9 cm long, style 6–9 mm long, stigma fused, 0.5–2 mm long. Fruit 10–33 mm long and 3–6 mm wide. Seeds with large recurved funicle. Fig. 4, p. 21.

Uganda. Mengo District: Mabira Forest, May 1908, *Brown* 439!
Tanzania. Kyela District: Mababu Village, 5 Mar. 2001, *Wiland & Mboya* 164!; Lindi District: Tendaguru, 12 Dec. 1930, *Migeod* 1023!
Distr. U 4; T 7, 8; Senegal, Gambia, Sierra Leone, Ivory Coast, Burkina Faso, Nigeria, Cameroon, Chad, Central African Republic, Congo-Brazzaville, Congo-Kinshasa; Madagascar
Hab. Grassland; 150–550 m

b. subsp. **minor** (*Guinea*) *Wiland* in Fragm. Flor. Geobot., 42(1): 16 (1997). Type: Equatorial Guinea, *Guinea* 399 (MA, holo.; BR!, iso.)

Plant up to 0.4 m high. Rhizome 1.4–11 cm long and 3–9 mm wide. Leaves 6–50 cm long and 1–5 mm wide; veins 5–27. Scapes 1–3, 1.2–4.7 cm long and 0.5–2 mm wide. Flowers solitary or rarely in 2–5-flowered racemes, hermaphrodite or male with a reduced pistil; lower bract 1.9–3.5 cm long and 3–5 mm wide, 11–15-veined; upper bract 1.3–1.5 cm long and 0.2–1.5 mm wide; in 2–5-flowered inflorescences each flower is subtended by only one bract, relatively wide and short, never fused in the basal part, 0.8–2.7 cm long and 1.3–4 mm wide, about 5-veined; perianth segments (6–)7–10(–12) mm long and (1.5–)2–2.5 mm wide; stamens 3–5 mm long, filaments 1.5–3 mm long, anthers 2–3 mm long, fused at the apex; ovary 5–10 mm long and 1–1.5 mm wide, rostrum 1.6–6 cm long, style 3–6 mm long, stigma fused or free, 0.5–2 mm long. Fruit 9–11 mm long, 3 mm wide. Seeds with a small recurved funicle, sometimes striate.

Uganda. Masaka District: Nabugabo, 5 Mar. 1933, *A.S. Thomas* 973!
Distr. U 4; Senegal, Sierra Leone, Liberia, Equatorial Guinea, Congo-Brazzaville, Congo-Kinshasa, Zambia
Hab. Swamp edge; 1300 m
Syn. *C. gallabatensis* sensu Durand & Schinz in Étud. Fl. État Indep. Congo: 260 (1896) pro parte, *non* Baker
    C. minor Guinea in Ensayo Geobot. Guin. Continent. Españ.: 258 (1946) & Ann. Jard. Bot. Madrid 6: 471 (1946); Hepper in F.W.T.A., 2 ed., 3: 174 (1968)
    Hypoxis angustifolia sensu Robyns & Tournay in Fl. Sperm. Parc. Nat. Albert 3: 208 (1955) pro parte, *non* Lamarck
    Curculigo pilosa sensu Geerinck in F.C.B.: 2 (1971) pro parte, *non pilosa sensu stricto*
    C. multiflora Zimudzi in Nord. J. Bot. 14: 311 (1994) & in Kirkia 16: 12 (1996); Nordal & Zimudzi in F.Z. 12(3): 4 (2001). Type: Zambia, Mwinilunga District, *Milne-Redhead* 2938 (K!, holo.)

c. subsp. **major** (*Bak.*) *Wiland* in Fragm. Flor. Geobot., 42(1): 19 (1997). Type: Nigeria, *Barter* 1506 (K!, holo.)

Plant up to 0.2 m high. Rhizome 2.5–7 cm long and 8–16 mm wide. Leaves linear, 12.5–20 cm long and 3–10 mm wide; veins 25–37. Scapes 1–5, 1-flowered, 1.7–2 cm long and 2–2.5 mm wide. Lower bract 4.8–7.1 cm long and 4–7 mm wide, 27–35-veined; upper bract 3.9–5.9 cm long and 1.5–3 mm wide, 9–13-veined; tepals 15–20 mm long and 3–5 mm wide; stamens 7–8 mm long, filaments 4.5–6 mm long, anthers 2.5–4 mm long, slightly emarginate; ovary about 19 mm long and 1.5 mm wide, rostrum 5.5–7.5 cm long, style 9.5–14.5 mm long, stigma fused, 0.5–1 mm long. Fruit 2.3–2.5 cm long and 4–7 mm wide. Seeds with a large, recurved funicle.

UGANDA. Karamoja District: N of Mt Kadam [Debasien], 4 May 1948, *Hedberg* 832!; Mengo District : Wabusana, Kalongo road, 14 Mar. 1933, *Lab. Staff T.* 1215!; Bunyoro District: Nord Chope, 2 May 1907, *Bagshawe* 1586!
KENYA. Kwale District: Shimba Hills, Longomwagandi, Mar. 1968, *Magogo & Glover* 469
TANZANIA. Mbeya District: Madibira–Ipogoro Road, 8 Dec. 1962, *Richards* 17344!; Masasi District: W of R. Bangala, 17 Dec. 1955, *Milne-Redhead & Taylor* 7704!
DISTR. **U** 1, 2, 4; **K** 7; **T** 7, 8; Senegal, Sierra Leone, Guinea, Burkina Faso, Ivory Coast, Ghana, Togo, Mali, Gambia, Nigeria, Niger, Chad, Sudan, Ethiopia, Cameroon, Central African Republic, Congo-Kinshasa, Burundi, Zambia, Malawi, Mozambique
HAB. Grassland, woodland; 350–1550 m

SYN. *Gethyllis pilosa* Schum. & Thonn. in Beskr. Guin. Pl.: 192 (1828); Baker in Trans. Linn. Soc., Bot. 2: 266 (1878), p.p. quoad *Barter* 1506
Curculigo gallabatensis Baker var. *major* Baker in J.L.S. 17: 123 (1878); Durand & Schinz, Cons. Fl. Afr. : 236 (1895)
C. gallabatensis sensu Chevalier in Expl. Bot. Afr. Occ. Fr.1: 635 (1920), *non* Baker
C. pilosa sensu Hutchinson & Dalziel in F.W.T.A. 2: 396 & Fig. 319 (1936) pro parte quoad *Barter* 1506

INDEX TO HYPOXIDACEAE

New names validated in this part

Hypoxis fischeri *Pax* var. **colliculata** (*Wiland*) *Wiland & Nordal* **comb. nov.**
Hypoxis fischeri *Pax* var. **katangensis** (*De Wild.*) *Wiland & Nordal* **comb. et stat. nov.**
Hypoxis fischeri *Pax* var. **hockii** (*De Wild.*) *Wiland & Nordal* **comb. et stat. nov.**
Hypoxis fischeri *Pax* var. **zernyi** (*Schulze*) *Wiland & Nordal* **comb. et stat. nov.**

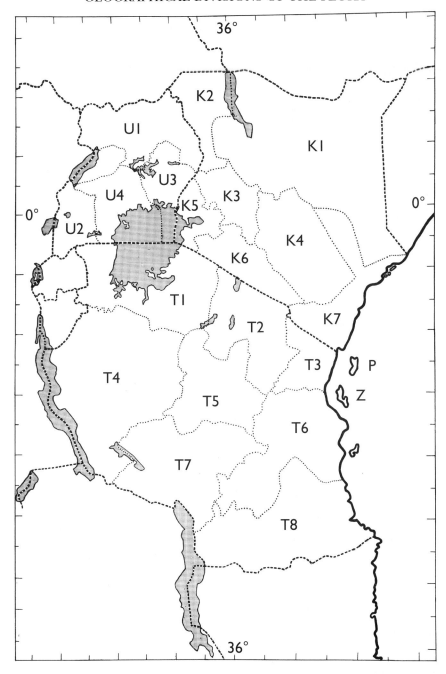

First published in 2006 by
Royal Botanic Gardens, Kew
Richmond, Surrey, TW9 3AB, UK
www.kew.org

ISBN 1 84246 167 2

British Library Cataloguing in Publication Data
A catalogue record for this book is available from the British Library

Design and typesetting by Margaret Newman,
Kew Publishing, Royal Botanic Gardens, Kew.

Printed in the UK by Hobbs the Printers

For information or to purchase all Kew titles please visit
www.kewbooks.com or email publishing@kew.org

All proceeds go to support Kew's work in saving the world's plants for life